SCIENTIFIC AMERICAN EDUCATIONAL PUBLISHING

SCIENTIFIC AMERICAN INVESTIGATES SPIDERS

GOLIATH BIRD-EATING SPIDERS

BY ROXANNE ALDOLPHO

I0811240

Published in 2025 by The Rosen Publishing Group
in association with Scientific American Educational Publishing
2544 Clinton Street, Buffalo NY 14224

Copyright © 2025 Rosen Publishing Group

Library of Congress Cataloging-in-Publication Data
Names: Aldolpho, Roxanne, author.
Title: Goliath bird-eating spiders / Roxanne Aldolpho.
Description: New York, New York : Scientific American Educational Publishing, an imprint of Rosen Publishing, [2025] | Series: Scientific American investigates spiders | Includes bibliographical references and index.
Identifiers: LCCN 2024006075 | ISBN 9781725351301 (library binding) | ISBN 9781725351295 (paperback) | ISBN 9781725351318 (ebook)
Subjects: LCSH: Theraphosa blondi–Juvenile literature. | Spiders–Juvenile literature.
Classification: LCC QL458.42.T5 A43 2025 | DDC 595.4/4–dc23/eng20240307
LC record available at https://lccn.loc.gov/2024006075

Portions of this work were originally authored by P. V. Knight and published as *Bird-Eating Spiders.* All new material in this edition is authored by Roxanne Aldolpho.

Designer: Andrea Davison-Bartolotta
Editor: Brianna Propis

Photo credits: Cover, p. 1 Wirestock Creators/Shutterstock.com; p. 5 THEPALMER/iStockphoto.com; p. 7 Dan Olsen/Shutterstock.com; p. 9 Griffin Photography23/Shutterstock.com; p. 11 blackeagleEMJ/Shutterstock.com; p. 13 ekler/Shutterstock.com; p. 15 Allmy/Shutterstock.com; p. 17 Bernard DUPONT/Flickr; p. 19 chiccosassi/iNaturalist; p. 21 Milan Zygmunt/Shutterstock.com.

All rights reserved. No part of this book may be reproduced in any form without permission in writing from the publisher, except by a reviewer.

Printed in the United States of America

CPSIA compliance information: Batch #CSSA25. For Further Information contact Rosen Publishing at 1-800-237-9932.

Find us on

CONTENTS

Words in the glossary appear in **bold** type the first time they are used in the text.

HAIRY AND SCARY!

It hides in **swamps** and under rocks in the **rainforests** of northern South America. Its name may send chills down a person's spine. The creature is covered with hair and creeps around on eight legs, killing its food with **venom**.

This scary-sounding animal is known as the goliath, or giant, bird-eating spider. It is a member of a group of large, hairy spiders called tarantulas. In fact, it's the biggest type of the group! But are these spiders as scary as they are hairy?

An adult goliath bird-eating spider is about the size of newborn puppy!

FUN FACT

THE GOLIATH BIRD-EATING SPIDER GOT ITS NAME AFTER EUROPEAN SETTLERS WITNESSED ONE EATING A HUMMINGBIRD.

AS HEAVY AS 30 QUARTERS

The goliath bird-eating spider is the second-largest spider in the world! With its legs spread out, it's 12 inches (30 cm) across—the size of a dinner plate! It weighs up to 6 ounces (175 g). That's equal to about 30 quarters!

The spider has two main body parts: the cephalothorax (seh-fuh-luh-THOHR-aks), which combines the head and **thorax**, and the abdomen, which contains the heart, guts, and more. At the end of the abdomen, parts called spinnerets produce silk. The spider's body is protected by a hard exoskeleton, or outside covering.

PARTS OF A GOLIATH BIRD-EATING SPIDER

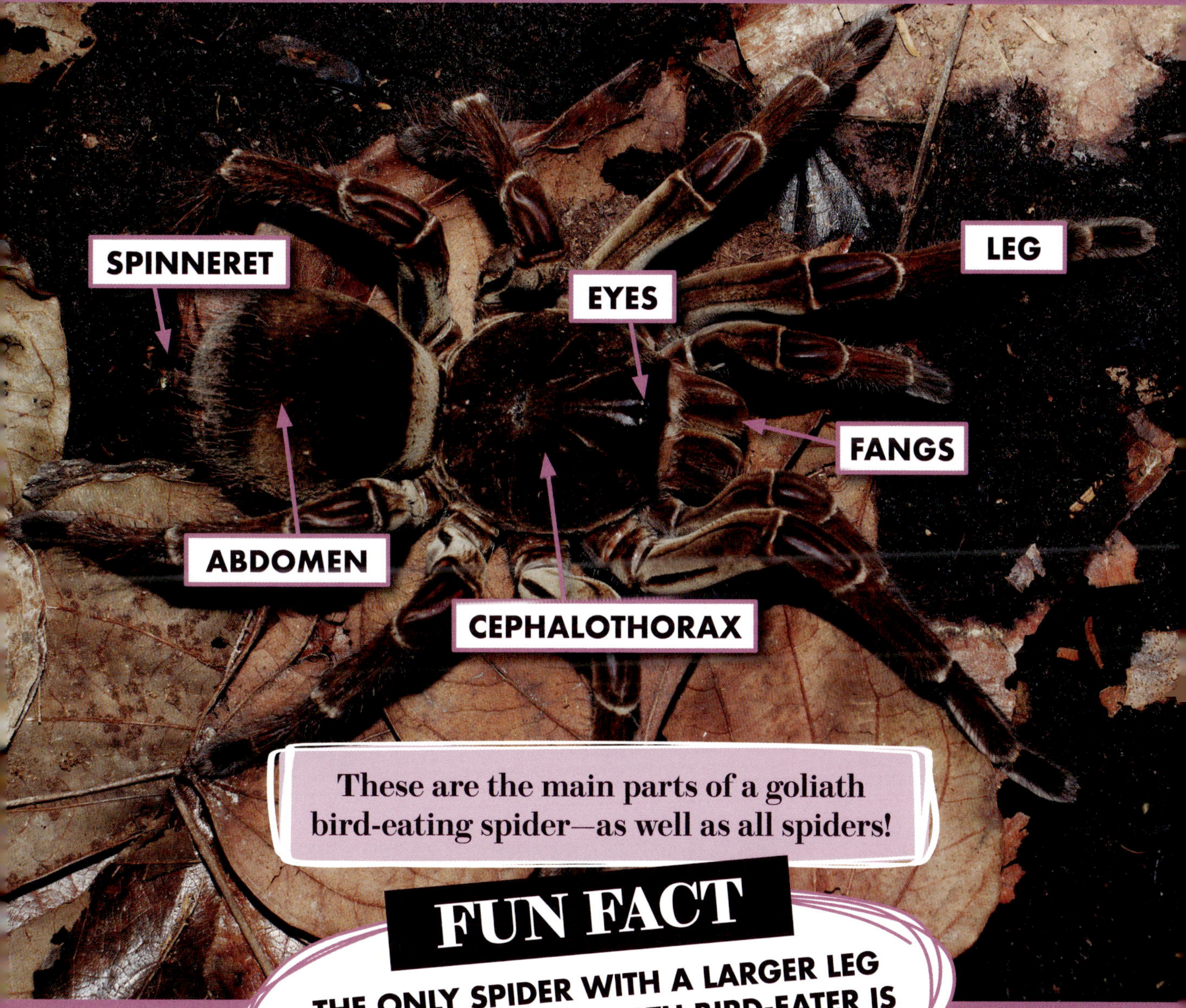

These are the main parts of a goliath bird-eating spider—as well as all spiders!

FUN FACT

THE ONLY SPIDER WITH A LARGER LEG SPAN THAN THE GOLIATH BIRD-EATER IS THE GIANT HUNTSMAN SPIDER.

VENOMOUS FANGS

The goliath bird-eating spider makes silk, but it doesn't spin webs to catch **prey**. So how does it find its food? It knows when a creature is nearby because the hairs covering its body sense the smallest movements in the air around it. Then it leaps suddenly and sinks its **fangs** into its prey! The venom **paralyzes** the prey, and the spider has a meal.

The goliath bird-eating spider eats bugs, frogs, small snakes, lizards, and other small animals. Although "bird-eating" is in the spider's name, it rarely feeds on birds!

The goliath bird-eating spiders' venom allows them to to eat anything smaller than they are.

FUN FACT

THE GOLIATH BIRD-EATING SPIDER HAS NO TEETH! ITS FANGS PUT DEADLY JUICES INTO ITS PREY THAT TURN THE PREY'S INSIDES INTO LIQUID. THE SPIDER SUCKS UP ITS MEAL WITH ITS STRAW-LIKE MOUTH.

HISSING OR SHOOTING?

If the goliath bird-eating spider finds itself in danger, it has many ways to protect itself from enemies. It may try to scare the creature off by rubbing together stiff hairs on its legs. This is called stridulation (strih-juh-LAY-shun). It makes a hissing noise loud enough to be heard 15 feet (4.6 m) away!

The spider can also shoot tiny hairs from its abdomen. These hairs stick in the enemy's skin and are highly **irritating**. The spider may also showcase its terrifying fangs by rearing up on its hind, or back, legs.

This goliath bird-eating spider is battling a scorpion! A scorpion has venom too, but it's found in the tail.

FUN FACT

THE ABDOMINAL HAIRS OF THE GOLIATH BIRD-EATING SPIDER CAN BE DEADLY. SMALL ANIMALS SUCH AS MICE CAN DIE FROM THE IRRITATION. THE HAIR FEELS LIKE SHARP PIECES OF GLASS!

RAINDROPS ARE FALLING ON THEIR HEAD

The goliath bird-eating spider can be found in the rainforests of northern South America, including Venezuela, Guyana, Suriname, French Guiana, and Brazil. It prefers to live deep in the rainforest, in the very drippiest, wettest parts, such as swamps.

The spider usually makes its home in a burrow, or hole in the ground, that it lines with silk. It may dig its own burrow or take over one left behind by another animal. It may also make its home under rocks or roots.

FUN FACT

LIKE MANY OTHER SPIDERS, THE GOLIATH BIRD-EATER IS SOLITARY—MEANING IT LIKES TO LIVE ALONE.

This map shows where goliath bird-eaters live in South America.

SPIDERLINGS

The life of a goliath bird-eater starts inside of an egg. A female lays 50 to 150 eggs in a large silk sack after **mating** with a male. She stores it in her burrow and guards it to keep it safe.

The babies, called spiderlings, come out of the eggs after about two months. They stay in their mother's burrow until they molt, or cast off their exoskeleton, for the first time. Afterwards, they leave the burrow and start their own lives.

FUN FACT

THE FEMALE MAY ADD HAIRS FROM HER ABDOMEN TO THE SILK SACK AROUND HER EGGS TO PROTECT THEM. THIS STOPS FLIES FROM HARMING HER UNBORN BABIES.

Goliath bird-eating spiders guard their eggs and also guard the entrance to their burrow if the eggs are inside.

GROWTH AND REGROWTH

As they grow into adults, spiderlings molt numerous times. It takes them two to three years to become adults. They continue to molt after they're grown. Because there's new growth every time they molt, they're able to grow a new leg if they've lost one!

Female goliath bird-eating spiders can live up to 20 years. Males usually live only three to six years. The reason: about half of all males are killed or hurt by females as they try to mate!

This is an old, cast-off exoskeleton. It looks like an actual goliath bird-eater!

FUN FACT

A GOLIATH BIRD-EATING SPIDER'S BODY IS SOFT AND RUBBERY AFTER MOLTING. THE NEW EXOSKELETON TAKES SEVERAL DAYS TO HARDEN, MEANING SMALL BUGS CAN EASILY KILL THE SPIDER DURING THIS TIME.

LET THEM BE

Although the mere sight of one could make some people shriek, a goliath bird-eater is not a danger to humans. It will only attack if it feels it's in danger. And most people say a bite is no worse than a wasp sting. Sometimes, however, it causes a lot of pain and makes you sick to your stomach.

Those irritating hairs the spider shoots from its abdomen can cause real problems if they get around your eyes and mouth. However, there is a simple solution to these problems: don't bother the spiders!

This large and hairy spider may be scary to look at! But remember: the goliath bird-eater doesn't want to hurt you.

FUN FACT

LIKE ALL SPIDERS, THE GOLIATH BIRD-EATER HELPS PEOPLE CONTROL THE BUG POPULATION.

FACING DANGERS

People are a big **threat** to goliath bird-eating spiders. For starters, many wild spiders are caught to sell as pets, causing an early death for the spiders. Secondly, the spiders' homes in the rainforest are being destroyed by **deforestation**. Lastly, people use **chemicals** to kill the bugs that the bird-eaters like to feast on, which harms the spiders. Some people in South America even find the spider tasty to eat!

The goliath-bird eating spider is a fascinating creature that helps people and their rainforest home. We need to keep these spiders safe!

It may seem like a cool idea to own a giant spider as a pet, but these creatures would rather live on their own in the wild.

FUN FACT

IT IS ESTIMATED THAT NINE OUT OF EVERY TEN SPIDERS CAUGHT TO BE PETS DIE BEFORE THEY CAN BE SOLD.

GLOSSARY

chemical: Matter that can be mixed with other matter to cause changes.

deforestation: The process of clearing an area of forest.

fang: A hard, sharp-pointed body part a spider uses to put venom in its prey.

irritating: Causing a stinging feeling.

mate: To come together to produce babies.

paralyze: To make unable to move.

prey: An animal hunted by other animals for food

rainforest: A forest that gets at least 100 inches (254 cm) of rain each year.

swamp: Wet, spongy land that has trees and is covered by water at least part of the year.

thorax: The part of the body between the head and the abdomen.

threat: Able to cause harm.

venom: Poison made in an animal's body and put into its prey by biting or stinging.

FOR MORE INFORMATION

Books

Higgins, Melissa. *Splendid Spiders.* North Mankato, MN: Pebble, 2020.

Lundgren, Julie K. *Spiders.* New York, NY: Crabtree Publishing Company, 2022.

Websites

National Geographic
https://www.nationalgeographic.com/animals/invertebrates/facts/goliath-birdeater
Learn even more about goliath bird-eating spiders, such as their diet and habitat.

Critter Stop
https://critterstop.com/post/tarantulas-as-pets-a-guide-to-caring-for-your-eight-legged-friend
Do you want a pet tarantula? Find out more about caring for one here.

Publisher's note to educators and parents: Our editors have carefully reviewed these websites to ensure that they are suitable for students. Many websites change frequently, however, and we cannot guarantee that a site's future contents will continue to meet our high standards of quality and educational value. Be advised that students should be closely supervised whenever they access the internet.

INDEX